Henry McGuier

A concise history of High Rock Spring

Henry McGuier

A concise history of High Rock Spring

ISBN/EAN: 9783337159986

Printed in Europe, USA, Canada, Australia, Japan

Cover: Foto ©ninafisch / pixelio.de

More available books at **www.hansebooks.com**

A

CONCISE HISTORY

OF

IGH ROCK SPRING.

By HENRY McGUIER.

SARATOGA SPRINGS:

STEAM PRESSES OF G. M. DAVISON.

1867.

High Rock Congress Spring Company.

OFFICERS.

JAMES M. MARVIN, - - - - - - - - - *President.*
JOHN H. WHITE, - - - - - - - - - - - *Secretary.*
JOHN S. LEAKE, - - - - - - - - - - *Treasurer.*
WILLIAM H. McCAFFREY, - - - - - *Superintendent.*

DIRECTORS.

JAMES M. MARVIN,	JOHN H. WHITE,
JOHN S. LEAKE,	WM. H. McCAFFREY,
SEYMOUR AINSWORTH,	DR. JOHN L. PERRY.

JOHN McB. DAVIDSON.

EXECUTIVE COMMITTEE.

JOHN H. WHITE,   JOHN S. LEAKE,
SEYMOUR AINSWORTH.

Laboratory of the School of Mines, Columbia College,

NEW YORK, Nov. 17, 1866.

WM. H. McCAFFREY, ESQ.,

Superintendent of High Rock Spring:

SIR: I have the honor to report the following results of the analysis of the water which I collected at the High Rock Spring, in Saratoga, in August last:

In one gallon of 231 cubic inches are contained,

Chloride of Sodium,	390.127	grains.
Chloride of Potassium,	8.974	"
Bromide of Sodium,	0.731	"
Iodide of Sodium,	0.036	"
Fluoride of Calcium,	trace.	
Sulphate of Potassa,	1.608	"
Bicarbonate of Baryta,	trace.	
Bicarbonate of Strontia,	trace.	
Bicarbonate of Lime,	131.739	"
Bicarbonate of Magnesia,	54.924	"
Bicarbonate of Soda,	34.888	"
Bicarbonate of Iron,	1.478	"
Phosphate of Lime,	trace.	
Alumina,	1.223	"
Silica,	2.260	"
Total,	628.039	grains.
Carbonic acid gas,	409.458	cubic inches.

Respectfully yours,

C. F. CHANDLER.

By reference to the analyses of the various other mineral fountains of Saratoga, and a comparison with the above, it will be readily seen that the water of the High Rock Spring is not only a much heavier water, but that it also contains a very much larger number of cubic inches of carbonic acid gas per gallon.

HIGH ROCK CONGRESS SPRING.

———•———

The proprietors of this fountain have adopted the following

TARIFF OF PRICES.

AT SARATOGA.		*AT NEW YORK.*	
Pints, per dozen,	$2 00	Pints, per dozen,	$2 25
Quarts " "	3 00	Quarts " "	3 50
Orders at Saratoga embracing one gross or more.		*Orders at New York embracing one gross or more.*	
Pints, per dozen,	$1 75	Pints, per dozen,	$2 00
Quarts " "	2 50	Quarts " "	3 00

Refilling Bottles.—Pints, 75 cents per dozen ; Quarts, $1.00 per dozen.

This water is put up in cases containing four, five and six dozen pints; and two, three and four dozen quarts.

Southern Depot for the sale of this water, No. 544 Broadway, New York, near Metropolitan Hotel,

NATURAL HISTORY

OF THE

MINERAL FOUNTAINS OF SARATOG⅃

————•—•••————

"Whence the origin of your mineral springs?"

This is a question often propounded to us by persons bo
of our own and other countries, who visit our world-fam
watering place. And, undoubtedly, to those who have r
made the complex operations of nature their study, there
very much of mystery connected with this matter. N
indeed, is there sufficient reason why our astonishment shou
be excited at this fact; for, amid the eternal activities of r
ture, and the illimitable resources from which she draws
largely for the production of her varied and ever varying pl
nomena, which "amid ceaseless changes seeks the unchangi
pole," the naturalist, alone, finds in the laborious study a
contemplation of those phenomena, a certainty which adm
of no question, and a reward, the munificence of which baff
the skill of the mathematician in his attempts at computatic
Nor, indeed, are any of her works so insignificant (if that
not a profanation) as not to demand his most serious and ca
ful consideration.

1*

That the reader may be enabled fully to understand the facts
bearing upon this subject, it seems necessary to state, succinctly, .
the geological character of this locality.

Immediately upon the north of the village of Saratoga
Springs, and within about three miles, we have the metamor-
phic rocks, (the *Taconic* system of EMMONS and the " *Quebec
group* " of the Canadian survey,) in mountainous ridges; trav-
ersed, diagonally by basaltic dikes, and at right angles, or nearly
so, by thin thread-like veins of a subsequently formed granite,
showing thereby, frequent and extensive occurrences of volcanic activity.

Superimposed upon its southern slope reposes the "Potsdam
sand stone" of Emmons, and No. 1 in the ascending series of the
New York and Silurean systems, with a dip to the south east,
varying from five to twenty degrees, and *in transitu*. Resting
upon the Potsdam sandstone is the "Calciferous sandrock" of
Eaton, No. 2 of the above systems, with a dip corresponding
with the preceding, as to amount and direction. This rock em-
braces, at this point, an area of about four square miles, and is
bounded on the south and east by the valley in which our mineral fountains are developed. The "Trenton limestone," No. 3,
(and by this designation I refer to the Chazy, Bird's eye, Black
river and Trenton limestones,) is well developed upon our
south west, about three miles distant, and loaded with its characteristic fossils, occupying a *horizontal* position and *in situ*.
Directly upon our south and east is developed the "Black " or
"Utica slate " of Vanuxem, No. 4 of the above systems, over-
laying the limestone, having its strata also *horizontal*.

Now, if we leave the railroad depot at this point, and pass
south along the line of the track for the distance of about two
miles, we shall pass from the *tilted up* surface of number two,
directly on to the *horizontal* surface of number *four*, *without
deviating perceptibly from a horizontal line*, thereby exhibiting
the existence of a *fault* or fracture in the rocks; and the tilted

up and fragmentary condition of the rocks upon the north of it, showing that they *must* have been broken off and thrust upward the whole thickness of numbers three and four, estimated at about half a mile.

It is conceded that if the strata of numbers two and four of this system, at this point, had the same dip throughout, it would be difficult to determine the existence of a fault, for the rocks *might* "over-lap;" but it occurs to find, in several localities, upon the north and west side of the valley, patches of the *lowest portion* of the Trenton limestone, superimposed upon the calciferous sandrock *and having the same dip*, but occupying a higher level than the *uppermost portion* of the same rock, just across the valley; while upon the opposite side of the fault, where it occurs *in place*, and is well developed, the strata are perfectly *horizontal;* a fact which *could not* exist had there been no displacement.

If this bird's-eye view of the geology of our locality be borne in mind, the reader will be enabled to comprehend the applicability, or otherwise, to our present subject, of conclusions based upon the existence of such FAULT.

Two theories have been advanced by which to explain or account for the origin of the mineral constituents of our springs, to wit:

1st. The solvent action of the water on the rocks, and their imbedded minerals, over or through which it passes.

2d. The sublimation or consolidation of the various gases thrown off by the internal fires of the earth, upon coming into contact with veins or bodies of pure water.

From the universally recognized solvent property of water, in its action upon the rocks, especially the calcareous and argillaceous, as well as upon many of their imbedded minerals, and its mechanical power of suspension exerted upon some of their constituents, the opinion has very generally obtained that most, if not all, mineral waters are produced in this manner;

and the more especially when, as it sometimes happens, a powerful chemical agent, in the form of a gas of some kind or
other, is present to aid in the production of such a result. Nor
has this opinion failed to find adherents among those who class
themselves with the scientific few. And perhaps, indeed, there
is some truth upon which to base such an opinion; for, from the
extensive dissemination of the various acids in a gaseous form,
and the greater facility with which water is thereby enabled to
erode the rocks subjected to its action, and the consequent
increased amount of mineral matter found in such waters,
seems to warrant something like such a conclusion. And yet
the *condition* under which such waters occur, must not be lost
sight of. The sulphur waters of the Hudson river valley, ex-
tending north and south a distance of at least 150 miles, with
a width of some 12 or 15 miles, developed in the "Black slate,"
(No. 4,) and depositing their sulphur upon coming into contact
with the atmosphere, will not, I think, be attributed to a similar
source as that from whence our mineral waters are derived;
for if we examine the slate we shall find it abundantly charged
with the *sulphuret of iron*, by which the phenomena of *these*
sulphur waters may be reasonably explained. But even in this
respect the rule is not persistent, either in relation to the origin
of the sulphureted hydrogen gas, or the mineral constituents
of waters charged with such gas.

Professor Lewis C. Beck, chemist of the New York State
Geological Survey, in speaking of the sulphur waters of Mon-
roe and Genesee counties, says: "To show how abundantly
sulphureted hydrogen is evolved in this district, it is only
necessary to notice the Caledonia springs in the town of Wheat-
land, where a large volume of water gushes out of the earth,
forming a stream nearly one quarter the size of the Genesee
river at Rochester, and so sour as to char the vegetable matter
over which it flows."

Of the various sulphur springs in Genesee county, he says of a single locality: "There is another locality of a similar kind a hundred rods west of Byron hotel, and two miles east of the former," (one previously noticed,) "which is remarkable, in consequence of the great quantity of acid. It is a spring which arises from the earth, in sufficient quantity to turn a light grist-mill. Such an immense laboratory of sulphuric acid is here conducted by nature, that the water which supplies this perennial stream possesses acidity enough to give the common test with violets, and to coagulate milk."

"It was my intention to have added to this general view of the sulphureted waters of our state, some remarks concerning the *origin* of the sulphureted hydrogen gas thus largely evolved; but I have only space to repeat, what has already been suggested, that the cause ordinarily assigned, viz: the decomposition of the sulphuret of iron, seems to me to be *wholly inadequate*, and that we must refer it to some agency far more general and effective."

An argument which we consider conclusive, is found in the fact that the sandrock, *number two*, and the black slate, *number four*, do not contain chloride of sodium, magnesia, soda or iodine, and especially iodine, which is *never* found in the rocks. And among the altered sediments of the "Quebec group," (7,500 feet thick,) no one has ever thought of looking for rock salt; consequently the advocates of the solvent theory will be driven to look somewhere else for those ingredients of our mineral waters. Nor is it true, so far as our observation extends, or upon the authority of others, that water passing over or through limerock ever liberates carbonic acid gas in quantities sufficient to exhibit its presence by its passage through, or escape from, such waters. And an important fact to be remembered in regard to the sulphur waters of the Hudson river valley, above alluded to, formed by solution, is, that they do not exhibit the least ebullition from the escape of the sulphur-

etted hydrogen gas, but which is sufficient to impart its odor
to the atmosphere for a considerable distance.

But we have actual demonstration; for within one hundred
and fifty yards of one of our most remarkable mineral springs
we have, *issuing from the same rock, of precisely the same tem-
perature*, a copious flow of pure fresh water. It therefore
remains for the advocates of the solvent theory to reconcile
this obvious "inconsistency with itself."

In examining the second theory, three questions very naturally suggest themselves, viz:

Are the saline ingredients found in our waters produced by
the process of sublimation of gases thrown off by the internal
fires of the earth, any where in nature?

Are waters ever charged with mineral constituents by such
process?

Are the conditions of the geological formations at Saratoga
favorable to the development of those elemental gases?

We find in our mineral waters chloride of sodium, forming
about one half their mineral constituents. Of which mineral
Humboldt says: "The vapors that rise from the *fumarolles*"
(small volcanic vents) "cause the sublimation of the chlorides
of iron, copper, lead and ammonium; iron glance and the
chloride of sodium (the latter often in large quantities) fill the
cavities of recent lava streams and the fissure of the crater."
(The chlorides of sodium, iron glance, sulphur, and indeed
some fifteen or twenty different metals and minerals are being
formed upon the inner surfaces of the cones of the craters of
Etna and Vesuvius, when not in a state of eruption, by the
sublimation or consolidation of the gases emitted by the internal fires of the earth, through those avenues. In fact the
rocks, and all the solid substances of the crust of the globe,
and even water itself, owe their origin to the process we are
now considering.) And hence we may very pertinently ask

whence the salt in sea-water? If we are answered, from solution; we reply that the *origin* of fossil salt is not suggested by such an answer; and because the cause is not adequate to the effect. It must be remembered that we have *thousands of cubic miles* of sea-water, and there are no known deposits of salt of sufficient capacity, in the aggregate, to supply such an enormous demand. But even if this were so, water, in no condition whatever, (if, as in the ocean, that condition be permanent,) can receive its saline ingredients by solution, and again deposit them; for the principle which causes the water to deposit the salt, would prevent it from taking it up in the first place. The Mediterranean is depositing crystals of salt on some parts of its bottom, at present. Lake Ooroomiah, in Persia, has deposited upon its bottom permanent alternating layers of salt and sand; a specimen of which water, examined in 1844, was found to contain about *one quarter part of solid salts*. The waters of Lake Elton, in Asiatic Russia, and other lakes adjoining the Caspian sea, have deposited thick beds of rock salt at their bottom. The same is true of Lake Indersk, on the steppes of Siberia. (*Daubeny on Thermal and Mineral Waters. Ure's Geology.*) And we find the process going forward in the great Salt Lake of Utah. Now, as solution is incompetent to impart to water the power to deposit salt, we are compelled to look to some other source for the salt contained in those waters which do deposit it at their bottom.

It may possibly be objected, that we have treated of a single one of the ingredients of our waters only. We reply, that all of the other constituents, even to iodine, are found in sea-water; and if the greater has its origin in the process of sublimation, it seems to us very rational to suppose that the less, existing in the same combination, should have its origin in the same cause.

In relation to the second inquiry, viz: Are waters ever charged with their mineral contents by the process of sublimation? We answer, unhesitatingly, yes; for it must be obvious to all, that if, as we have shown is true, those gases *do* sublimate or resolve themselves into solid compounds upon coming into contact with the atmosphere, they most assuredly will upon coming into contact with a denser medium. But happily for us, this position is not unsupported by very high authority. President Hitchcock, in quoting Prof. Daubeny upon this subject, says: "When these springs" (thermal—which is the character of our springs) "occur in volcanic districts, their origin is very obvious. The water which percolates into the crevices of the strata becomes heated by the volcanic furnace below, and impregnated with salts and gases by the sublimation of matter from the same focus."

Dr. Daubeny has shown that "thermal springs *not* in volcanic districts, in a large majority of cases, rise either from the vicinity of some *uplifted chain of mountains* or from *clefts* and *fissures caused by the disruption of the strata;* and are, therefore, in all cases, probably the result of deep-seated volcanic agency, which may have long been in a quiescent state."

Humboldt says, "We see issue from the ground, steam and gaseous carbonic acid, carbureted hydrogen gas and sulphurous vapors. Such effusions from the *fissures* of the earth not only occur in districts of still burning or long extinct volcanoes, but they may likewise be observed occasionally in districts where neither trachyte or any other volcanic rocks are exposed on the earth's surface. * * * We see in Germany, in the deep valley of the Eifel, in the neighborhood of the lake of Laach, in the crater-like valley of the Wehr and in western Bohemia, exhalations of carbonic acid gas manifest themselves as the last efforts of volcanic activity in or near the foci of an earlier world."

Now, if we revert to the geologic epitome presented by us in the outset, we shall discover that we are in the *immediate* "vicinity of some uplifted chain of mountains;" that Nos. 1 and 2 of the New York and Silurean systems are made to assume a parallelism with the southern slope of the mountains upon our north, i. e. having a dip of about 20 degrees, and in some instances, as at the Empire Spring, of full 45 degrees, and which necessarily implies a disruption of the strata, unless they were in a plastic state at the time of the application of the disturbing force; an idea readily dissipated, when we remember that Nos. 3 and 4 of the above systems occupy a *horizontal position*, and that the surface of No. 2 is on a level with the surface of No. 4, thereby clearly indicating the fact that Nos. 1 and 2 have not only been broken off but actually thrust upward, the entire thickness of Nos. 3 and 4 of our system; and unless there is a wide vacuity between the *Plutonic* rocks on the one hand, and the lower sedimentary rocks on the other, (an impossibility,) this *fault* or *fissure* extends, necessarily, to the internal fires of the earth, and all the conditions competent to explain the phenomena of our mineral waters by the method we are now considering is, in my judgment, fully established.

And here we submit, that as an avenue is opened at this point, (as we have already abundantly shown,) through which the gases from the internal fires of the earth can escape, which are *now* producing chloride of sodium (common salt) in other localities, that it is quite as philosophical, to say the least, to attribute its existence in our mineral waters to that cause as to any secondary source; and the more especially, as such secondary source *can not be shown to exist in this vicinity;* unless, indeed, it be demonstrated that the operations of nature are not persistent.

And in this connection (we may say without incurring the charge of egotism) it is certainly gratifying to know that the views above expressed have received the endorsement of Prof. JOSEPH HENRY of the Smithsonian Institute, not only by placing a manuscript copy of the same in the archives of that institution, but also by suggesting the idea of giving it publicity in this popular form.

Origin and Age of High Rock.

The material of which this rock is composed is principally impure lime, and is chiefly derived by the water from the loose earthy materials laying upon the rock out of which it issues. This material is quite different from any thing originally found in the water, and is retained in it by a mechanical instead of a chemical force, and consequently, upon its coming into contact with the atmosphere, and losing much of its activity, it deposits all those materials which have combined with it in its passage from the rocky orifice to the surface, in the form of a stony mass, denominated *tufa*. This is the origin, and such the substance forming that singular phenomenon known as the "High Rock."

In all the operations of Nature, everywhere, she has left the evidences of some method by which to determine the successive stages of progressive development and perfection, in all her varied creations. The geologist finds, in the rocks, unquestionable evidences of the stately steppings of the creative energy, and by their organic reliquæ or imbedded petrifactions is enabled to determine the comparative remoteness or nearness of the system he is studying. So, too, the botanist finds in the towering giant of the forest the annular rings of its growth, and he is thereby enabled to trace its history far backward, and perhaps prior to the commencement of his own brief existence. And the paleontologist, by comparing one specimen with another, is enabled to determine the mature from those which are immature; and so throughout.

The application of this law, then, to any subject of natural history to which our attention may be called, will enable us to arrive, approximately at least, at the truth, whenever we endeavor to trace backward to the commencement of their operations, those causes which have been instrumental in producing it.

Taking this law for our guide, then, let us determine, if possible, the age of the HIGH ROCK.

In descending from the surface at this point, seven feet of commingled muck and tufa (rocky matter formed by the water) was passed through, then a stratum or layer of tufa two feet thick, a stratum of muck, and then a stratum of tufa three feet thick.

In determining the time requisite to deposit the five feet of tufa, I caused a specimen of the tufa to be ground down smooth, and at right angles to the lines of deposit, so as to be enabled to count the lines, with accuracy, of annual deposit—as the vicissitudes of our climate determine those lines, for when frozen, as in our winters, the water makes no deposit—I found twenty-five such lines embraced within a single inch, and as there are sixty inches in the aggregate, a very simple computation shows that one thousand five hundred years were consumed in depositing these layers of tufa alone; and this tufa, it must be remembered, was deposited from standing water, or with but very little motion, as the tufa occupies a horizontal position.

Laying upon the stratum of tufa three feet thick, and in the stratum of muck superimposed upon it, was found a pine tree, the annular rings of which I counted to the number of one hundred and thirty; this sum added to the above, and we have the further sum of one thousand six hundred and thirty years. And from the foregoing data I deem it a moderate

approximation to claim four hundred years as the requisite
time in which to deposit the seven feet of superincumbent muck
and tufa, which [gives the still further sum of two thousand
and thirty years.

The facts which add strength to the foregoing conclusions,
and lend thrilling interest to this subject, are the evidences
which are found, at this depth from the surface, that this level
was once occupied by human beings. Here the extinguished
fire, marks unmistakably the gathering place of the family
group, many centuries ago. And here, too, linger the "foot
prints" of a long gone race, as if loth to leave a spot once so
cherished, and around which clustered so many pleasing recollections.

The reader will observe that the above estimate does not
include the rock or cone of the spring, but simply the *intermediate* strata between the cone and the deposits below. To determine the length of time requisite to form the cone or rock
of the spring, it became necessary to visit a locality where the
water, which is now depositing tufa, has a velocity similar to
that which the water *must* have had from which the rock of
the High Rock Spring was deposited. Accordingly, resort was
had to such a locality, and it was found that five of the annual
strata thus deposited occupied the space of one sixteenth of an
inch—thus requiring eighty years to perfect one inch; and as
the cone of the High Rock is four feet in height, it must have
required three thousand eight hundred and forty years to have
formed the cone. And in the aggregate, five thousand eight
hundred and seventy years (some eminent scientists who have
had their attention drawn to this subject, estimate its age at
even more than this,) must have been consumed in the formation of the HIGH ROCK SPRING.

Chronology of High Rock.

Away down amid the unnumbered decades of centuries, embosomed in the depths of a primeval forest, whose stillness was unbroken save by the stealthy tread of Nature's own sons, or the flocks which she had so munificently provided for them; in a valley of surpassing wildness and beauty; in the land of a republic the most beneficent, perfect and enduring the world ever saw, (and but for the destructive advance of the pale-faced invaders would have been perpetual,) the Great Spirit Sire, in view of the wants of the brave, guileless, magnanimous red man, smote for him the rock, and, at the omnipotent behest, up leaped the fountain, limpid as the "kohinoor," and *more* priceless than the golden wedge of Ophir. No wonder, then, that the red child of nature,

———— "whose untutored mind
Sees God in clouds and hears him in the wind,"

should bring hither his sick ones, or meet in annual convocation to pay his devotions, from a heart unsullied by guile and uncontaminated by the cold hypocrisy of later times. And no wonder that the Great Spirit Father, pleased with his offering, should determine to embellish with a vase of incomparable beauty and symmetry, the red man's pool of SILOAM.

Having glanced, in the preceding pages, at the natural history of the mineral springs of Saratoga, we shall now attempt a chronological history of the great, and indeed only spring, at

2*

this point, known to the inhabitants, whether savage or civilized, for long periods of time ; and the only one for which nature ever prepared and garnished with her own hands, a channel through which it might be presented to the invalid sufferer
without invoking any artistic interference whatever. Hence
nature has indicated this one as her favorite *jet d'eau;* her
chosen *alma mater.* The only one upon which she has left her
own unerring, enduring impress.

Its history, running through several centuries, is replete
with stirring events ; surrounded by mythical legends, gar-
landed with oriental metaphors, and embellished with all the
high wrought fiction so characteristic of the Six Nations. It
tells of battles fought, and won, and lost. It tells of a "proud
and powerful republic;" its commencement, its growth, its
advantages, and its strength. It tells of levees, held annually,
around the pool of "sweet waters" to please the Great Spirit.
And, alas! it also tells of the broken-hearted red man wrap-
ping his blanket around him, taking his last, sad, farewell look
at the spring which the Great Spirit gave him, and departing,
dejectedly and forever, to other hunting grounds, far away
towards sundown.

It is unquestionably true that centuries ago, and long before
any of the Caucassian race ever dreamed that such a continent
as the American had been thrust up from beneath the waters
of the turbulent Atlantic, or in fact existed any where, the
aborigines congregated around the High Rock fountain and
appropriated to themselves the advantages which it proffered.
The evidences which exist confirmatory of this view, although
not numerous, are most striking and decisive.

Beneath the surface of the valley in which this fountain is
situated, as it exists to-day, and at the depth of about twelve
feet, was discovered an ancient fire-place. The filling up of
the intermediate space (by natural processes) between it and
the present surface, could not have consumed less than two

thousand years; and if to this we add the time requisite to produce the rock or cone, as we now find it, we shall have more than five thousand years of intervening time between the period when the builders of that ancient fire-place sported in festive glee, or practiced their epicurean skill, or celebrated their victories of the chase and the foray, or, perhaps, planned their deeds of warfare and aggression, around this time-honored fountain; and by the light and heat of this primeval representative of a modern palatial hotel, fared sumptuously upon the avails of the field and the war-path, and the time when it came to the hands of its present proprietors.

Human advancement, all experience tells us, is extremely tardy, and this is true, whether in the *Stone Age*, or in the present age—the *Iron Age*. In the implements which the red man has left, scattered profusely all over this region, we find evidences of the rudest condition of the race, and also of a high degree of advancement upon that condition.

From implements of the most rude and uncouth character ever required or used by uncultured men, to the most perfectly finished axe, hatchet, javelin, amulets, personal ornaments, war-club with its nicely carved wolf's head, and various domestic utensils, made from stone, so hard that with all our boasted superiority over the red race, we are still unable to comprehend by what process they performed feats of skill which confound and bewilder our most astute lapidaries; and in this respect exhibiting a lapse of time quite as astonishing as that presented by the geological indications above referred to.

The language, too, of the ancient inhabitants of this spot, which was but an almost unintelligible jargon once, has arisen, through long periods of time, to the dignity of written signs as expressive of ideas, and hence we find many of their more finished implements bearing evident markings of characters giving tangibility to thoughts, or recording the progress of

time; and thus on until it stands out in the full proportions
of a written history.

Having thus emerged from what has generally been considered the dim, shadowy and unreliable domain of legend and
tradition, we now approach the more certain and reliable his-
toric region of civilized life; and·essay to cull and write out
its teachings in relation to a subject which has excited the
attention and admiration of the world.

At the beginning of the 18th century the red man was in
quiet, peaceful possession of that portion of the domain of the
Six Nations, known in aftertimes as the "Patent of Kayaderosseras." In 1703 the authorities, under the British crown,
gave permission to certain persons to purchase from the Mohawks, one of the tribes of the republic known as the Six
Nations, the tract of country of which that patent or grant is
composed. In 1704 the title was perfected. So secretly was
this title obtained, and so quietly held, that many years
elapsed before the entire nation of Mohawks (Mohocks) became aware of their loss. Upon its discovery, many were the
complaints made to Sir William Johnson, until at length in
August, 1768, a meeting of the agents of the patentees and
the chiefs of the Mohocks took place. "The Mohocks, [says
Sir William Johnson,] who, on examining the deed and survey,
and receiving a handsome sum of money, were at length *pre-
vailed* on to yield their claim to the patentees in my presence."

Just one year previous to the occurrence above narrated, on
a beautiful day in August, four stalwart Indians might have
been seen bearing upon their shoulders a litter, upon which
was reposing an invalid pale face—the friend of the red man—
in the person of Sir William Johnson, the cortege headed by
one McDonald. (He stated this fact to Mr. G. M. Davison in
1819.) They had brought their enfeebled "olive tree" to their

own Siloam, and he was healed—but for them it was a costly offering upon the altar of friendship.

From this time forward, the fame of this wonderful natural production spread over the land, and another incentive was presented to stimulate the cupidity of the white race to make still farther aggressions upon the home of the "poor Indian," which aggressions have continued until at last no single representative of that once proud and powerful people remains as an occupant of their once happy homes; but they either roam as wanderers or are gathered to their fathers to occupy more peaceful hunting grounds upon which no aggression is permitted.

On Friday, February 22, 1771, the patent of Kayaderosseras was partitioned, by ballot. And lot number twelve of the sixteenth general allotment, on which lot the High Rock Spring is situated, by such balloting, came into the possesion of Rip Van Dam. This is the first individual white man who ever exercised any possessory jurisdiction over this spring. Dying soon after, his executors sold the same to Isaac Low, Jacob Walton and Anthony Van Dam. Low was attainted for treason by the legislature of New York. October 1, 1779, and Henry Livingston, upon the sale of Low's portion of the lot, purchased the same for himself and several of his brothers. The property or lot was again divided in 1793. At this time it was held by Henry Walton, Henry Livingston and Anthony Van Dam. Walton then purchased Van Dam's portion of the property.

In 1826, Dr. John Clarke, the then proprietor of Congress Spring, through fear that in the hands of its then proprietors, or some other persons, the High Rock Spring, (which was then the only spring at Saratoga of any note save the Congress,) might undergo improvements which would enable it to become a successful rival of the Congress, purchased that portion of it belonging to the Livingstons. This portion, by his death in May, 1846, descended to his widow and heirs.

In the same year, Mr. John H. White, a step-son of Dr. Clarke, on behalf of Mrs. Clarke and the heirs, purchased of the executor of Henry Walton the remaining portion of the High Rock, and they thus became possessed of the entire property.

In 1864, William B. White, who succeeded Dr. Clarke in the control and management of the Congress Spring, died, and soon after it passed into other hands, and the *necessity* for the longer retention of this, to them entirely unproductive property ceased to exist; and in 1865 Messrs. Ainsworth and McCaffrey became the owners of this prodigy of nature.

These gentlemen soon after commenced a series of improvements which have resulted most advantageously to themselves and the fountain. After removing the building which sheltered the spring they set about removing the rock or cone, whole, upon accomplishing which, contrary to general expectation, they discovered that the cone had no direct or immediate connection with the rock below, but that the water was supplied by percolation through the intervening soil. They at once determined upon removing the soil quite down to the permanent orifice in the rock below, and by supplying an artificial channel between that point and the surface, to re-produce that much desired spectacle of the water once again bubbling up and running over the crest of the cone. After passing through about seven feet of commingled muck and tufa, they came upon a layer of tufa about two feet thick, then a stratum of muck, then another stratum of tufa three feet thick; through the muck were disseminated the trunks of large trees and pine and other forest leaves, in profuse abundance—the concentric rings of the trunk of one of those trees I counted, and found *one hundred and thirty* — those trees must have lain there for a long period of time before they became covered by the increasing peaty deposit, for their upper surfaces were worn smooth by the moccasins of the Indians, as they formed a convenient passage-way for them to the spring; and thus proceeding through alternating strata of muck and tufa down to the desired point, where an opening was reached which furnished a volume of water vastly superior to any thing ever before witnessed at this place, and so great, even, as to affect materially, for the time, the level of the springs in the neighborhood, some of them to the extent of quite two feet; thus exhibiting the fact that this is the main opening of all our mineral waters at this point. A tube was then furnished, placed in position and properly secured, in which the mineral water rose several feet above the original surface of the rock or cone. Preparations were immediately made for replacing the rock back upon the vein of water, and after considerable labor and trial that purpose was accomplished, and water welled up through the orifice and overflowed the rock; a spectacle never before presented to the admiring gaze of a white man.

Since then, as if by magic, there has started into existence a towering and capacious building, drest in the Italian style of architecture, designed for, and adapted to, the economic pur-. poses of the proprietors of the spring. But the fountain itself has received an embellishment, which to be properly appreciated must be seen. A pavilion within a pavilion. The style of its architecture is that of the gothic, and most admirably adapted and proportioned in all its parts; the whole surmounted by a mosque-like dome which adds much to the exquisite beauty of the finish. The dome itself is surmounted by an emblem significant of the jealous care with which this fountain has ever been regarded, whether in the possession of the red or the white man.

This point having been arrived at, and all the necessary preparations completed to reproduce the overflow of the waters of the fountain, it was suggested that this was an appropriate occasion for a general convocation to witness and celebrate the event by the white man, as, in the long past, it was the practice of the red man. Accordingly, on the 23d of

August, 1866, (the same month in which the Six Nations used
to hold their annual levees here,) a national salute ushered in
the day, and the busy note of preparation betokened the approaching ceremonies. At 1 o'clock the venerable WALWORTH,
president of the day, with STONE, the orator, and invited guests,
appeared in the forum. And citizens, and strangers from every
part of the country, gathered in throngs, crowding the building and grounds to overflowing, to listen to the orations and to
catch a glimpse of a phenomenon never before vouchsafed to a
white man. The following report of the proceedings is copied
from the *Daily Saratogian:*

After the speakers and invited guests had ascended the
staging erected for their use, Chancellor WALWORTH, president
of the day, called the assemblage to order, and delivered the
following interesting address:

LADIES AND GENTLEMEN: We are assembled at this time to celebrate the successful achievement of two of the enterprising
citizens of this town, Messrs. AINSWORTH and McCAFFREY. They
have taken up this renowned " High Rock" from the argillaceous
bed upon which it had probably rested for centuries, have ex-
plored the hidden aperture in the calciferous sand stone below
the clay, through which aperture it received its healing waters,
and have again restored it to its place; where, I trust, it is de-
stined long to remain, the wonder, as well as the pride of Saratoga. And, what is of far more importance to us, and to the
people of the United States generally, they have, by excluding
the fresh water from this very ancient fountain of health, doubled
the mineral strength of the medicinal waters of the High Rock
Spring, and have thereby greatly improved their healing properties. And this spring is from this time to take its proper place
as the oldest, and as one of the brightest of the stars in that
splendid galaxy of sparkling medicinal fountains which have
already made Saratoga the most celebrated, as well as the best
watering place in the world.

Many have supposed this rock to be of recent origin. And some assert that the healing waters have flowed out of its top subsequently to the commencement of the American Revolution. But all of them are unquestionably wrong. The top of this rock arose five or six feet at least above the highest point of the bed of clay upon which the rock had been formed by the gradual deposit of the mineral substances which had been chemically combined with the water; which water ebbed and flowed, at short intervals, as you see it does now. And geologists will tell you that it required a very long time to form a rock of that height, by such gradual accretions, before the water ceased to deposit new particles of mineral matter by flowing over the top of this rock. This High Rock Spring has been known to white men as a medicinal fountain, for about one hundred years; and perhaps longer. Sir William Johnson, who lived at Johnstown, about forty miles to the west of it, and who died in July, 1774, was brought here by the Indians a few years before his death, to partake of its healing waters. In the fall of 1777, after the surrender of General Burgoyne, and while our troops lay at Palmertown, about six miles north of here, several of our officers visited this spring, which had then attained some celebrity, as one of those officers has since told me. And it had for a long time before that been known to the Indians as "The Great Medicine Spring."

When the mineral waters of this ancient spring, which are this day, (by artificial means,) made again to flow over the top of this rock, ceased to flow over, is not known to any one now living. But I will give you the information I have on that subject. I first visited Saratoga in the summer of 1812, fifty-four years since. The water in this rock was then about as much below the top of the rock as it was when I came here to reside, eleven years afterwards, I think eighteen or twenty inches, or perhaps a little more. The late Major-General Mooers, of Plattsburgh, who was an officer of Colonel Hazen's regiment, at the taking of General Burgoyne's army, was at my house, and visited this spring with me, a few years previous to his death. He then told me that he, with other officers, came from Palmertown to this spring, in October,

3

1777. And he said the height of the water in the rock was then about the same as it was when we visited it, sixty years thereafter.

About forty-one years since, while holding a circuit court on the northern frontier of this state, I stayed over the Sabbath with a friend who resided a few miles from the Indian settlement at St. Regis; and we attended the religious services at the Indian church in their village. Between the morning and afternoon services at the church, we went to the house of one of their chiefs, named Loran Tarbel, with whom I had become acquainted during my residence at Plattsburgh. He was then between eighty and ninety years of age, but was in health and in perfect mental vigor. Knowing that some of the St. Regis Indians had once resided on the banks of the Mohawk river, I was anxious to learn what this aged chief knew in relation to this spring. But as he had a very imperfect knowledge of the English language, I spoke to his son, Captain Tarbel, who had an English education. I described the High Rock Spring, and asked him if he knew any thing about it. He said he had never been here, and had never heard of it. I then requested him to describe it to his father, and to ask him if he had ever heard of it. The moment he did so the early recollections of the venerable chief were aroused; and indicating by the motions of his hand the shape of the top of the rock, he said, " Yes, Great Medicine Spring."

He then told me, through his son as interpreter, that he was born at Caughnawaga, on the Mohawk; and that he emigrated with his father to Canada several years before the revolutionary war. That, when he was a boy, the Indians living on the Mohawk were in the habit of visiting this spring and using its waters as a medicine. That when he was about fifteen years old, and shortly before he emigrated to Canada, he came here with his father to see the Great Medicine Spring. I then asked him if the water flowed over the top of the rock at that time. He said it did not; that they had to get the medicine water by dipping it out of the rock with a cup or gourd shell. That there was then a tradition among the Indians that the medicine water had formerly flowed out of the rock at its top, but that it had ceased to 'do so for a long time before he came here with his father. He then gave me the Indian tradition as to the cause of the cessation of the overflowing of the water. The particulars of this tradition I can not repeat, in his words, in [the presence of this audience; but the substance of it was that the Great Spirit, who had made this wonderful rock, and had caused the 'healing waters to flow from it spontaneously for the benefit of his red children, was angry on account of the desecration of its medicine waters in making so improper use of them by some of their squaws, who had visited the spring and taht the water never flowed over the rock afterwards.

Such was the tradition of the untutored Indians, who knew little of geology or of hydraulics. But the true reason why the mineral waters ceased to flow out at the top of this rock, which had been gradually formed from their deposits, was probably this: These waters, in process of time, had found another outlet, perhaps at some considerable distance from here, and which outlet must have been something like twenty inches lower than the level of the top of this rock. For we now see that by tubing the mineral fountain so that it can not escape from beneath, or in any other way than through this natural orifice at the top of the rock, the present proprietors of the spring now cause its healing waters to flow out again, where they had ceased to flow for more than a century at the least.

As the enterprise of these proprietors has thus secured the control of these waters, and has greatly improved their medicinal value, it is of but little importance whether the water is hereafter to be permitted to flow over the rock into artificial basins, or is to be drawn from within or from beancath it, or by other means, for public or private use.

The whitening of the head of him who now addresses you, by the snows of seventy-eight winters which have fallen upon it, admonishes him to recollect that he can enjoy with you this valuable addition to our health preserving mineral fountains, only for a very short period. Still I rejoice with you all at the success of this enterprise, because I believe it will greatly benefit others and be a source of health and enjoyment to the people of every section of our beloved country. I fervently pray, therefore, that the healing and health preserving waters of this now renovated spring, may long continue to flow from this time honored rock, or be drawn from it or beneath it, to benefit and bless my fellow men. And as the civil war, which has recently scourged this once happy country, has now terminated by the submission of those who attempted to separate states whose union the founders of the constitution had declared perpetual, I hope and trust that Saratoga hereafter may continue to be, as it was a few years since, a common center of attraction, where all the people of the glorious union, who desire to come hither for health or pleasure, can meet together as brothers and sisters of a common family, without disturbance from the withering curse of sectional agitation. For the God of Heaven and earth has decreed

> That never again shall our country have slaves,
> "While the earth bears a plant, or the sea rolls its waves."

May we all remember that our Saviour has told us the blessing of Heaven rests upon the promoters of peace and good will among men, as contradistinguished from those who sow the seeds of discord and fan the flames of strife. And may the glory and the felicity of the re-united states of this great confederacy, over all of which the Star Spangled Banner now waves triumphant, continue to increase with each revolving year, until the thundering Cotopaxi shall cease to burn, and the cloud-capped Chimborazo be sunk in the ocean.

At the conclusion of the Chancellor's address the band struck up " The Star Spangled Banner," after which the chairman introduced the Orator of the day, WILLIAM L. STONE, Esqr., of New York, who pronounced the following eloquent oration.

" What song the Sirens sang, or what name Achilles assumed when he hid himself among women, though puzzling questions, are not beyond all conjecture. What time the persons of these ossuaries entered the famous nations of the dead, and slept with princes and counsellors, might admit a wide solution. But who were the proprietors of these bones, or what bodies these ashes made up, were a question not to be resolved by man, nor easily, perhaps, by spirits." Thus discoursed Sir Thomas Browne in his fearful essay upon Urn Burial, which he was led to write by the discovery of the celebrated urn in a "field of Walsingham" more than two hundred years ago. Fortunately for the day and the occasion, no such mystery now hangs over the wonderful masterpeice of nature, whose clustering memories we are here this day to recall. This spot, nevertheless, is consecrated ground. We may here, figuratively at least, tread upon the dust of kings. How long were their line and their triumphs we can not tell. Farther back than three hundred and fifty years, history herself is silent. Beyond that time America herself was "one great antiquity" buried in darkness of five thousand years.

I have said that we were treading upon the ashes of kings. It is indeed a fact that the royal title was unknown in their own imperfect language. But in their rank, their order of descent and their manner of exercising power, they were sovereigns and their chief sachems kings. "Rude kings they were, it is true. Kings who reveled not in voluptuousness, nor wasted their time amid the delights of the harem, nor degraded their manhood by plying the distaff like Sardanapolus. Nor yet were they of those who sought immortality by rearing cities and palaces and solemn temples, like those of Thebes and Babylon and Tyre. They affected not the graves of giants, nor yet sought to mark the age of their glory by the stupendous pyramid or the costly mausoleum." They were not of the common order of men, but a race proud and haughty — whose persons and characteristics were of mingled grandeur and gloom — and who, like the Fates of Grecian mythology, seemed born amid the convulsion of the elements, in cloud and storm. It is to this kingly race that we owe the priceless boon of the spring now before us. Help me, then, to lift with reverend hands the veil that, until now, has shrouded it in mystery.

Recent investigations have established the fact that the medicinal properties of the "High Rock" were well known to the

3*

Iroquois Confederacy fully two hundred years before the prows of Jacques Cartier's vessels, in 1535, grounded upon the emerald shores of the St. Lawrence. It was called by them, as my venerable and learned friend, who has just preceded me, has said, the " Medicine Spring of the Great Spirit," under whose special guardianship it was supposed to be. There can be no question, moreover, that the water flowed over the rock during this period, since there is yet a well authenticated tradition that it was only when the Great Spirit had been seriously offended by one of the Mohawk tribe, that he manifested his displeasure by causing its flow to cease. Reticent, however, as the Indian race naturally are, the discovery of America had been made many years before it was first brought to the notice of the whites; and it is probable that it would have remained unknown for many years longer, but for a most remarkable series of events, which, under the guidance of an overruling Providence, brought its properties into notice, and gave to the New World a Pool of Bethesda, for the healing of the halt, the lame, and the infirm. To show the wonderful manner in which this was brought about, is the object of him who now addresses you.

The " High Rock Spring " is deserving of more than the ordinary interest that attaches to the springs of Saratoga, not only on account of its being the greatest mineral curiosity on the globe, and of the superior character of its water, but because it was the first spring known to the whites in America.

The first man who visited it was Sir William Johnson, Bart. Sir William, under a commission of Major General from his Majesty, George II, defeated the flower of the French army, under Baron Dieskau, at the battle of Lake George, on the 8th of September, 1755. In this action he received a severe wound by a bullet in his thigh, from the effects of which he never wholly recovered, but was frequently subject to serious illness. At such times the wound, from which the ball was never extracted, became excessively painful, rendering him for weeks, after an attack, unable to ride on horseback or to endure any active exercise. Suitable medical attendance it was very difficult to procure,

and it frequently happened that having exhausted the contents
of his own medical chest, he was obliged to send to Albany, and
sometimes to New York, for a physician. It was during one of
these attacks, in the summer of 1767, that the Mohawks determined, in solemn council, to reveal to their beloved brother, War-
ra-ghi-ya-ghy, the peculiar medicinal properties of the " High
Rock." Nor, perhaps, could there have been any stronger proof
of the affection in which he was held by these sons of the forest,
than their resolution to give their brother the benefits of that
which they had always sacredly guarded as the precious gift to
themselves alone, from the Great Spirit. Accompanied by his
Indian guides, the Baronet set out on his journey the 22d of
August, and passing down the Mohawk from Johnstown in a boat,
soon reached Schenectady. At this place, being too feeble either
to walk or ride, he was placed on a litter and borne on the stalwart shoulders of his Indian attendants through the woods to
Ballston lake, which he reached the same evening. Tarrying
over night at the log cabin of Michael McDonald, a Scotchman,
who had recently begun a clearing in the vicinity, the party,
three hours before sunrise, on Thursday, the 23d of August,
1767, plunged again into the forest; and following the trail of
Indian hunters, along that which is now the road from Ballston
to this village, came to the chief tributary of Lake Saratoga, the
Kayaderosseras.

In the gray dawn of that summer morning, along the green
aisles of the primeval forest, the party silently pursued their
way. The moccasined feet pressed down the wild flowers in
their path. Wheeling above with untiring wing, as if moving with
and watching over the party, were several noble bald eagles, whose
eyries hung on the beetling crags, affording to the invalid a pre-
sage of health and happiness. Aloft the pine tree towered above
a sea of verdure, and below, the maple, whose virgin cheeks were
not yet brazen with the paint of early frosts, modestly shrunk
from the passing gaze. " Old fir trees hoary and grim, shaggy
with pendant mosses, leaned above the stream," and beneath,

32   HISTORY OF

dead and submerged, a fallen sycamore thrust from the current the bare, bleached limbs of its collossal skeleton.

The sun was an hour above the eastern hills, when the startled deer saw the evergreens sway, and the Baronet's party emerge from the thicket. Their polished bracelets and rich trappings, glittering in the dewy foliage like so many diamonds, were in keeping with the cheerfulness visible upon each countenance — for were they not bearing their dearly beloved brother to. the medicine spring of the Great Spirit? As the party emerge from the glade upon the green sward, they separate into two divisions, and, with gentle tread, approach the spring, bearing their precious burden in the center. Pausing a few rods from the spring, the Baronet leaves the litter; and, for a moment, his manly form, wrapped in his scarlet blanket bordered with gold lace, stands towering and erect above the waving plumes of his Mohawk braves. Then, approaching the spring, he kneels, with uncovered head, and reverently places upon the rock a roll of fragrant tobacco — his propitiatory offering to the Manitou of the spring. Still kneeling, he fills and lights the great calumet, which, through a long line of kings, had descended to the renowned Pontiac, and taking a whiff from its hieroglyphic stem, passes it to each chieftain in turn. Then, amid the profound silence of his warriors, he for the first time touches his lips to the water; and gathering the folds of his mantle about him, amid a wild and strange chant raised by the Indians to their Deity, he enters the rude bark lodge which, with prudent forethought, his braves had erected for his comfort, directly where this building now stands; and in this *primitive hotel* reclined the *first* white man that had ever visited this spring. Yet while the sufferer lay on his evergreen couch, did the fortunes of the General whom he had defeated twelve years previously, occur to him? Perhaps so; for by a singular coincidence, while the conqueror of Dieskau was prostrated amid these forests, where the wounds of both had been received, the French General was languishing on his deathbed at a small town in the interior of France —

" The paths of glory lead but to the grave."

The Baronet had been but for four days at the "High Rock," when he received letters obliging him to hasten immediately home. Short as his visit was, however, the water restored his strength so far as to enable him to travel some of the way to Schenectady on foot; and again taking his water carriage, he arrived on the 4th of September, at the Hall, to welcome his son Sir John, who had just arrived from England. The popularity of Saratoga Springs, as a watering place, may be said to date from this visit. "My dear Schuyler," writes the Baronet upon his return, to his intimate personal friend General Phillip Schuyler, "I have just returned from a visit to a most amazing spring, which almost effected my cure; and I have sent for Dr. Stringer, of New York, to come up at once and analyze it." Accordingly when Schuyler effected a settlement on the banks of the Hudson, it was, undoubtedly, the remembrance of this letter that caused him, in 1783, to cut a road of twelve miles through the forest to this spring and erect a tent, under which himself and his family spent several weeks, using the water. Hence it was that the fact of so distinguished a personage as Sir William having been restored by the water soon became noised through the country. Others were induced to make the trial; new springs were discovered; and thenceforth the springs became the resort of those who were in pursuit of health and pleasure. For many years after its discovery, the "High Rock" continued to be the resort of people from all sections of the country; and when other springs were found in the neighboring village of Ballston, the chief drive of the visitors was a romantic drive through the woods to the "High Rock Spring." The question will here very naturally be asked, if the High Rock was so celebrated, how did it happen that it has remained until this late day comparatively unknown?

The answer is very simple. In April, 1826, the late proprietors of "Congress Spring" bought the former from Walton and Livingston, and kept it in the background. These proprietors, however, having recently died, the spring was purchased from the heirs by its present owners, Ainsworth and McCaffrey. The object of these gentlemen in making the purchase was to bring the High Rock into such prominence before the public as its real value as a restorative demanded. Accordingly, no sooner had the sale been completed, than — as *Reconstructions* are all the order of the day — they resolved to "reconstruct" their purchase, and endeavor, if possible, to cause the water again to overflow the rock. The project at first, as with "Reconstructions" generally, did not meet with public favor, as fears were entertained that Saratoga might be deprived of one of her greatest attractions. But in the face of numerous obstacles they persevered, and the result — of which you have visible proof to-day — has demonstrated the practicability of their plan, and all the mysteries of the High Rock and its spring have been unveiled to the public gaze. A slight excavation showed that the rock only extended a few inches below the surface, and it was easily removed. Within it was a chamber about two feet in diameter, and below a pit formed by the bubbling water, about ten feet in depth, in which were found a large number of tumblers lost in dipping the water. Around the cone, for an area of four hundred feet, the soil was found to be filled with two independent layers of encrustations or *tufa* — formed by the deposits of the water — one of them three feet in thickness, and the other two. Immediately beneath the rock lay the body of a pine tree, eighteen inches in diameter, which still retained its form, and was sufficiently firm to be sawn in sections and pulled out. This tree must have fallen before the formation of the surface rock commenced, and had undoubtedly lain there thousands of years. For many years before the stalagmite formation of the cone hid it from sight, this tree — evidently placed there by design — was used as a convenient pathway to the spring, since the upper side of the log has been worn to a polished surface by the moccasins of the aboriginals.

A very interesting question here arises. What is the age of this remarkable fountain? The rock itself was formed, as you doubtless are aware, by the precipitation of minerals held in solution by carbonic acid gas. The rock or cone is four feet in height. Now by counting the annual deposits of *tufa*, it is found that five of the yearly layers measure one sixteenth of an inch.

Hence eighty years are required to deposit a single inch, or nine hundred and sixty years for a foot. From this it appears that the age of the rock from its first formation to the period when the water — having been forced by hydrostatic pressure into another outlet — ceased to overflow the rock, can not be less than four thousand years. And if to this be added the time consumed in forming the *tufa*, which is two thousand years more, we have six thousand years, as near as geological investigation can determine, as the age of this mineral fountain itself — placed here by the Almighty two centuries before he created man in his own image — while darkness yet brooded over the face of the deep. The excavation was continued about twelve feet, when it became evident that only a few inches more would bring to view the crevice in the solid rock out of which this wonderful fountain unceasingly flows. The tubing is now fitted to the rock, so as to exclude all extraneous substances and confine the gases ; and it is confidently believed that a superior mineral water has been obtained, which will be available for commercial purposes.

Thus is it that in the hands of its present owners the ambrosial nectar of the gods becomes a veritable fact, and the " elixir of life," sought for so many years in vain by the alchemists of old, finds in this spring its realization. Upon this Rock, Hebe may break her cup, and, chagrined and discomforted, acknowledge that her vocation is at an end. Had the " High Rock Spring " stood on the borders of the Logo d' Agnaus, the noted Grotto del Cani would never have been heard of beyond the environs of Naples ; while this fountain in its place would have been deservedly celebrated in story, to the admiration of the world, as one of the greatest of curiosities !

It remains only to speak of the agency which the battle of Lake George exercised in bringing this spring into notice. Indeed, the parallel that exists between the benefits which that action conferred upon our national and physical life is so striking, that a brief glance at it may not be omitted by those who read the hand of God in every event of life. The action of the 8th of September, 1755, so far as concerns the number of men engaged, was not a great battle; but when viewed in its immediate strategical results, it well deserves a prominent place among the battles of American history. "The battle of Lake George," says the late Reverend Cortlandt Van Rensselaer, in his admirable discourse upon this action, "is memorable in defeating a well-laid, dangerous scheme of the enemy, and in saving the provinces from scenes of bloodshed and desolation. If Dieskau had succeeded in overthrowing Johnson in his intrenchments, his advance upon Fort Edward would have been easily successful, and thence his march to Albany would have been triumphant. The conflagration of our northern settlements would have been followed by the desolation of Albany and Schenecady; and although Dieskau must have soon been compelled to retreat, it is impossible to estimate the bloodshed, plunder and general losses, which might have taken place had not God ordered it otherwise. The victory of Lake George undoubtedly rescued the province from injury and woe beyond computation; considered, therefore, in its immediate strategical results, the battle was one of the most important engagements in American history. The battle of Lake George is also remarkable for its influence in rallying the spirit of the American colonies. Much had been expected from the three expeditions sent against the French; but disappointment and sorrow had already followed Braddock's terrible defeat. All the provinces were amazed, awe struck for a time, but recovering from the first shock of the calamity, they were aroused to avenge their loss.

Johnson's victory was received as the precursor of a recovered military position and fame, and was hailed as a means of deliverance from a bold and cruel foe. Few battles ever produced more immediate results in rekindling military and martial enthusiasm. Congratulations poured in upon General Johnson from every quarter. Not only were the colonies filled with rejoicing, but the influence of the triumph went over to England, and the deeds of our fathers at Lake George became familiar to the ears of royalty, and were applauded by the eloquence of Parliament."

But again. The battle of Lake George was furthermore

memorable in its suggestions of provincial prowess, and its lessons
of warfare to the colonies preparatory to their independence.
It is a mistake to suppose that Bunker Hill was the first school in
which the colonists were taught their ability to struggle with
veteran soldiers. It was at Lake George that this lesson was
learned; and it is very doubtful whether the colonists would
have dared to have taken the stand they did, had it not been for
the lessons of the old French war. The battle was fought by provincial troops, and chiefly by the sons of glorious old New England. The veteran regulars of old England had been beaten in
the forests of Western Pennsylvania, or remained inactive in the
Niagara expedition. Through some unaccountable cause, the
expedition, which was on the direct line of Canada, and nearest
to the French reinforcements known to be at hand, was con-
signed exclusively to the care of native colonial soldiers; and
bravely did they do their duty. On these shores provincial
prowess signalized its self-relying capabilities; and in this battle
and in this war the colonists practically learned the value of
union. Putnam, and Stark, and Pomeroy came here as to a
military academy to acquire the art of warfare, which they
all exercised at Bunker Hill. George Washington himself, as a
military man, was nurtured for America and the world amid the
forests of the Alleghanies, and the rifles and tomakawks of these
French and Indian struggles, Lake George and Saratoga, are
continuous not merely in territory, but in heroic association.

As this battle, therefore, was in a measure the source of our
present national life, so, by leading indirectly to the discovery of
this pring, it has been a source of renewed physical energy to the
nation. One is but the correlative of the other. *Sana mens in
corpore sano* is as true of the body politic, as of the body
physical; and if our existence as a nation is preserved, it
will be by keeping intact the mental and physical energies
of the people. "Soldiers," said Napoleon, on the eve of one
of his battles, and in one of those bulletins with which he
was wont to electrify all Europe. "Soldiers, from yonder
pyramids, forty centuries are gazing down upon you!" But on

the eve of the battle of Lake George, from far nobler and grander
heights the Providence of God was looking down moulding and
shaping its results for the benefit of mankind throughout the
ages. "And he showed me a pure river of water of life,
clear as crystal, proceeding out of the throne of God and of the
Lamb. In the midst of the street of it, and on either side of
the river, was there the tree of life, which bare twelve manner
of fruits, and yielded her fruit every month; and the leaves of
the tree were for the healing of the nations."

Aware of the importance attached, by the public, to every
thing pertaining to this heir loom which has come down to us
from an epoch much earlier than our own, and the apparent
mystery which hangs about the whole; I have been thus particular in collating and presenting the chronological facts relative
to this subject. Those obtained from the Chief of the Tusca-
roras are perfectly authentic, and, so far as I am aware, are
entirely new to the civilized world.

The white man's chronology, too, of this particular wonder
of the world, has hitherto existed only in the form of musty
title deeds, running through long periods of time, and held by
individuals residing in different and distant sections of the
country. And it is confidently believed that this is the first
time in which a connected and continuous history of this remarkable fountain has ever been presented to the public.

www.ingramcontent.com/pod-product-compliance
Lightning Source LLC
Chambersburg PA
CBHW021448090426
42739CB00009B/1686